Title: ____________________ Project: ______________

Continued from:

To Page No:

Signed	Date	Witnessed	Date

Book: Page: **2** Title: ____________________ Project: ____________________

From Page No:

To Page No:

Signed	Date	Witnessed	Date

From Page No:

To Page No:

Signed	Date	Witnessed	Date

Book: Page: **4** Title: ____________ Project: ____________

From Page No:

To Page No:

Signed	Date	Witnessed	Date

Title: ____________________ Project: ___________________________ Book: Page: **5**

From Page No:

To Page No:

Signed	Date	Witnessed	Date

Book: Page: **6** Title: ____________ Project: ____________

From Page No:

To Page No:

Signed	Date	Witnessed	Date

Title: ____________________ Project: ____________________________ Book: Page: **7**

From Page No:

To Page No:

Signed	Date	Witnessed	Date

'Follow and Save' (See back page)

Book: Page: **8** Title: ____________ Project: ____________

From Page No:

To Page No:

Signed	Date	Witnessed	Date

Title: ____________________ Project: ____________________________ Book:

From Page No:

To Page No:

Signed	Date	Witnessed	Date

Book: Page: **10** Title: ____________ Project: ____________

From Page No:

To Page No:

Signed	Date	Witnessed	Date

From Page No:

To Page No:

Signed	Date	Witnessed	Date

Book: Page: **12** Title: ____________________ Project: ____________________________

From Page No:

To Page No:

Signed	Date	Witnessed	Date

From Page No:

To Page No:

Signed	Date	Witnessed	Date

Book: Page: **14** Title: ____________________ Project: ____________________

From Page No:

To Page No:

Signed	Date	Witnessed	Date

Title: ____________________ Project: ____________________________ Book:

From Page No:

To Page No:

Signed	Date	Witnessed	Date

Book: Page: **16** Title: ____________ Project: ____________

From Page No:

To Page No:

Signed	Date	Witnessed	Date

Title: ____________________ Project: ____________________________ Book:

From Page No:

To Page No:

Signed	Date	Witnessed	Date

Book: Page: **18** Title: ____________ Project: ____________

From Page No:

To Page No:

Signed	Date	Witnessed	Date

Title: ____________________ Project: ____________________________ Book: Page: **19**

From Page No:

To Page No:

Signed	Date	Witnessed	Date

'Follow and Save' (See back page)

Book: Page: **20** Title: ____________ Project: ____________

From Page No:

To Page No:

Signed	Date	Witnessed	Date

Title: ____________________ Project: ____________________________ Book:

From Page No:

To Page No:

Signed	Date	Witnessed	Date

Book: Page: **22** Title: ____________ Project: ____________

From Page No:

To Page No:

Signed	Date	Witnessed	Date

Title: ____________________ Project: ____________________________ Book:

From Page No:

To Page No:

Signed	Date	Witnessed	Date

Book: Page: **24** Title: ____________________ Project: ____________________

From Page No:

To Page No:

Signed	Date	Witnessed	Date

Title: ____________________ Project: ____________________________ Book:

From Page No:

To Page No:

Signed	Date	Witnessed	Date

Book: Page: **26** Title: ____________ Project: ____________

From Page No:

To Page No:

Signed	Date	Witnessed	Date

Title: ____________________ Project: ____________________________ Book:

From Page No:

To Page No:

Signed	Date	Witnessed	Date

Book: Page: **28** Title: ____________ Project: ____________

From Page No:

To Page No:

Signed | Date | Witnessed | Date

Title: ____________________ Project: ____________________________ Book:

From Page No:

To Page No:

Signed	Date	Witnessed	Date

Book: Page: **30** Title: ______________ Project: ______________

From Page No:

To Page No:

Signed	Date	Witnessed	Date

From Page No:

To Page No:

Signed	Date	Witnessed	Date

Book: Page: **32** Title: ____________ Project: ____________

From Page No:

To Page No:

Signed | Date | Witnessed | Date

Title: ____________________ Project: ____________________________ Book:

From Page No:

To Page No:

Signed	Date	Witnessed	Date

Book: Page: **34** Title: ____________ Project: ____________

From Page No:

To Page No:

Signed | Date | Witnessed | Date

From Page No:

To Page No:

Signed	Date	Witnessed	Date

Book: Page: **36** Title: ____________________ Project: ____________________________

From Page No:

To Page No:

Signed	Date	Witnessed	Date

From Page No:

To Page No:

Signed	Date	Witnessed	Date

Book: Page: **38** Title: ____________________ Project: ____________________________

From Page No:

To Page No:

Signed	Date	Witnessed	Date

From Page No:

To Page No:

Signed	Date	Witnessed	Date

Book: Page: **40** Title: ____________________ Project: ____________________________

From Page No:

To Page No:

Signed	Date	Witnessed	Date

Title: ____________________ Project: ____________________________ Book:

From Page No:

To Page No:

Signed	Date	Witnessed	Date

 Title: ______________ Project: ______________

From Page No:

To Page No:

Signed	Date	Witnessed	Date

Title: ____________________ Project: ____________________________ Book:

From Page No:

To Page No:

Signed	Date	Witnessed	Date

Book: Page: **44** Title: ____________ Project: ____________

From Page No:

To Page No:

Signed | Date | Witnessed | Date

From Page No:

To Page No:

Signed	Date	Witnessed	Date

Book: Page: **46** Title: ____________ Project: ____________

From Page No:

To Page No:

Signed	Date	Witnessed	Date

Title: ______________________ Project: ______________________________ Book:

From Page No:

To Page No:

Signed	Date	Witnessed	Date

Book: Page: **48** Title: ____________ Project: ____________

From Page No:

To Page No:

Signed	Date	Witnessed	Date

From Page No:

Signed	Date	Witnessed	To Page No: Date

Book: Page: **50** Title: ______________ Project: ______________

From Page No:

To Page No:

Signed	Date	Witnessed	Date

From Page No:

To Page No:

Signed	Date	Witnessed	Date

From Page No:

To Page No:

Signed | Date | Witnessed | Date

Title: ____________________ Project: ____________________________ Book: Page: **53**

From Page No:

To Page No:

Signed	Date	Witnessed	Date

Book: Page: **54** Title: ____________________ Project: ____________________________

From Page No:

To Page No:

Signed	Date	Witnessed	Date

Title: ____________________ Project: ____________________________ Book:

From Page No:

To Page No:

Signed	Date	Witnessed	Date

Book: Page: **56** Title: ____________ Project: ____________

From Page No:

To Page No:

Signed | Date | Witnessed | Date

From Page No:

To Page No:

Signed	Date	Witnessed	Date

Book: Page: **58** Title: ____________ Project: ____________

From Page No:

To Page No:

Signed	Date	Witnessed	Date

Title: ____________ Project: ________________ Book:

From Page No:

To Page No:

Signed	Date	Witnessed	Date

Book: Page: **60** Title: ____________ Project: ____________

From Page No:

To Page No:

Signed	Date	Witnessed	Date

From Page No:

To Page No:

Signed	Date	Witnessed	Date

Book: Page: **62** Title: ____________ Project: ____________

From Page No:

To Page No:

Signed | Date | Witnessed | Date

Title: ____________ Project: ________________ Book: Page: **63**

From Page No:

To Page No:

Signed	Date	Witnessed	Date

Book: Page: **64** Title: ____________ Project: ____________

From Page No:

To Page No:

Signed	Date	Witnessed	Date

From Page No:

To Page No:

Signed	Date	Witnessed	Date

Book: Page: **66** Title: ____________________ Project: ____________________________

From Page No:

To Page No:

Signed	Date	Witnessed	Date

From Page No:

To Page No:

Signed	Date	Witnessed	Date

Book: Page: **68** Title: ____________________ Project: ____________________

From Page No:

To Page No:

Signed	Date	Witnessed	Date

Title: ____________________ Project: ____________________________ Book: Page: **69**

From Page No:

To Page No:

Signed	Date	Witnessed	Date

From Page No:

To Page No:

Signed	Date	Witnessed	Date

From Page No:

To Page No:

Signed	Date	Witnessed	Date

Book: Page: **72** Title: ____________ Project: ____________

From Page No:

To Page No:

Signed	Date	Witnessed	Date

From Page No:

To Page No:

Signed	Date	Witnessed	Date

 Title: ____ Project: ____

From Page No:

To Page No:

Signed | Date | Witnessed | Date

From Page No:

To Page No:

Signed	Date	Witnessed	Date

Book: Page: **76** Title: ____________ Project: ____________

From Page No:

To Page No:

Signed | Date | Witnessed | Date

From Page No:

To Page No:

Signed	Date	Witnessed	Date

Book: Page: **78** Title: ____________________ Project: ____________________

From Page No:

To Page No:

Signed | Date | Witnessed | Date

Title: ______________________ Project: ______________________________ Book:

From Page No:

To Page No:

Signed	Date	Witnessed	Date

Book: Page: **80** Title: ____________________ Project: ____________________

From Page No:

To Page No:

Signed	Date	Witnessed	Date

From Page No:

To Page No:

Signed	Date	Witnessed	Date

Book: Page: **82** Title: ______________ Project: ______________

From Page No:

To Page No:

Signed | Date | Witnessed | Date

Title: ____________ Project: ________________ Book:

From Page No:

To Page No:

Signed	Date	Witnessed	Date

Book: Page: **84** Title: ____________ Project: ____________

From Page No:

To Page No:

Signed	Date	Witnessed	Date

Title: ____________ Project: ____________ Book:

From Page No:

To Page No:

Signed	Date	Witnessed	Date

Book: Page: **86** Title: ____________________ Project: ____________________________

From Page No:

To Page No:

Signed	Date	Witnessed	Date

Title: ______________________ Project: ______________________________ Book:

From Page No:

To Page No:

Signed	Date	Witnessed	Date

Book: Page: **88** Title: ____________ Project: ____________

From Page No:

To Page No:

Signed	Date	Witnessed	Date

From Page No:

To Page No:

Signed	Date	Witnessed	Date

Book: Page: **90** Title: ____________ Project: ____________

From Page No:

To Page No:

Signed	Date	Witnessed	Date

From Page No:

To Page No:

Signed	Date	Witnessed	Date

Book: Page: **92** Title: ____________ Project: ____________

From Page No:

To Page No:

Signed	Date	Witnessed	Date

Title: ____________ Project: ____________ Book: Page: **93**

From Page No:

To Page No:

Signed	Date	Witnessed	Date

Book: Page: **94** Title: ____________ Project: ____________

From Page No:

To Page No:

Signed	Date	Witnessed	Date

Title: ____________________ Project: ______________________________ Book: Page: **95**

From Page No:

To Page No:

Signed	Date	Witnessed	Date

Book: Page: **96** Title: ____________ Project: ____________

From Page No:

To Page No:

Signed	Date	Witnessed	Date

From Page No:

To Page No:

Signed | Date | Witnessed | Date

From Page No:

To Page No:

Signed | Date | Witnessed | Date

Title: ____________ Project: ________________ Book:

From Page No:

To Page No:

Signed	Date	Witnessed	Date

Book: Page: **100** Title: ________ Project: ________

From Page No:

To Page No:

Signed	Date	Witnessed	Date

Title: ____________________ Project: ____________________________ Book: Page: **101**

From Page No:

Continue to:

Signed	Date	Witnessed	Date

Unit Conversion Tables

Linear Measure

1 inch	= 25.4 millimeters	1 millimeter	= 0.039 inch
1 foot = 12 inches	= 0.3048 meter	1 centimeter = 10 mm	= 0.394 inch
1 yard = 3 feet	= 0.9144 meter	1 decimeter = 10 cm	= 3.94 inches
1 (statute) mile = 1,760 yards	= 1.609 kilometers	1 meter = 100 cm	= 1.094 yards
		1 kilometer = 1,000 m	= 0.6214 mile

Square Measure

1 square inch	= 6.45 sq. centimeters
1 square foot = 144 sq. inches	= 9.29 sq. decimeters
1 square yard = 9 sq. feet	= 0.836 sq. meter
1 acre = 4,840 sq. yards	= 0.405 hectare
1 square mile = 640 acres	= 259 hectares

Capacity (American Liquid)

1 pint = 16 fluid ounces	= 0.473 liter
1 quart = 2 pints	= 0.946 liter
1 gallon = 4 quarts	= 3.785 liters

Weight

1 pound = 16 ounces	= 0.4536 kilogram	1 kilogram = 1,000 grams	= 2.205 pounds
1 stone = 14 pounds	= 6.35 kilograms	1 tonne (metric) = 1,000 kilograms	= 0.984 (long) ton
1 quarter = 2 stones	= 12.70 kilograms		
1 hundredweight = 4 quarters	= 50.8 kilograms		
1 (long) ton = 20 hundredweight	= 1.016 tonnes		
1 short ton = 2,000 pounds	= 0.907 tonne		

Made in the USA
Middletown, DE
24 August 2024